I0476621

Introduction to Non-transitive Gambling Bets for Magicians
Written and Compiled by Bruce Carlley
Copyright 2015

Table of Contents

Table of Contents

What is this book about?

I am going to show you a number of games that the magician or gambler plays against a spectator one on one. The spectator gets a free choice of the objects that are used in the gambling game. (Some games use cards, dice, pieces of paper, spinners, etc.) For example, say the spectator has a choice of four dice. Spectator picks one die and the magician picks one die. Each rolls their die, and the high number wins. Whoever wins a best of 12 game match wins the game.

Unfortunately for the spectator, the magician will almost always win the match. The probability will ALWAYS be in the magician's favor. There is no sleight of hand or funny business. These type of games are called non transitive games.

What is a non transitive game?

A beats B. B beats C. And C beats A!. The best explanation would be to give an example that you may know. Rock-Paper-Scissors. The game is played between 2 people. Each player puts one hand behind their back and forms one of three shapes with their hidden hand. A rock (closed fist), paper (a flat hand) or scissors (index and ring fingers forming a "v" shape. When someone says "go", they simultaneously bring out their hands. Rock beats scissors by crushing it. Scissors beats paper by cutting it. Paper beats rock by covering it. If both players show the same shape, it is a tie. Usually the game is played to decide who will buy lunch, pay for drinks, etc. There is no known winning strategy if each player chose his shape randomly. There are computer programs and people that can beat you if you use some type of detectable pattern. Also some people will change their shape in mid "throw" if they see what your shape might be.

Here is a truly fair game of rock, paper, scissors. Each person is given 11 pieces of paper. They each write down either rock, paper, or scissors on each piece of paper in secret. A person could write rock on 6 pieces of paper, and scissors on the other 5, paper on all 11 pieces , or any other combination on the papers. When ready to play, person A puts down the 11 pieces of paper face down in a row. Person B puts his pieces of paper next to person A's. Now each pair is turned face up and the result recorded. If person A's paper has rock and Person B has scissors, B wins one point. Whoever has more wins in the 11 pairs is the winner.

Is it possible to cheat in this method? Opaque paper will prevent seeing whats on the other side. Perhaps a magician's assistant can cue the number of rocks, paper and scissors the spectator wrote. The magician can then adjust the quantity of his own symbols. For a simple example, if the assistant cues that the spectator wrote "rock" on all 11 slips of paper, then the magician can write "paper" on at least 6 slips of paper to insure the win. Also a hidden camera can show the assistant what is written. In any event, besides cheating, there is no mathematical way (That I know of!) that can give you a true edge. In all the games that follow, you will always have a true mathematical edge without cheating.

Letter Roulette by Royal Magic

Description from the box:

  They spin, you win!
  The magician and a spectator each select several letters, and spin the miniature roulette wheel to arrive at a combination of numbers and letters. The magician almost always wins - even when playing for low total. Better than loaded dice because the numbers and letters appear to be random. However, the operator controls the game and can even pick his own odds. Sold for entertainment purposes only.
It is still sold at some magic shops as Alphabet Roulette.

  All true for this disguised pseudo non-transitive game. The game consists of a card with 16 letters on it corresponding to the letters on the roulette game. The roulette wheel consists of two wheels. The outer wheel has 16 numbers and the inner wheel has 16 letters.

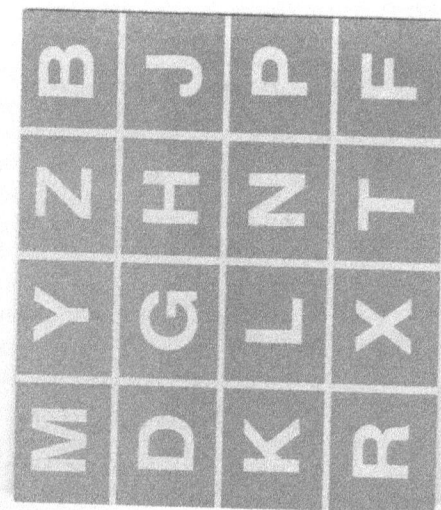

To play, the spectator selects a letter on the bingo card and puts a coin on it. The magician then selects a letter and puts a coin on it to mark it. The inner roulette wheel is spun. Whatever selected letter lands on the higher number wins.

For example, the spectator selects the letter "G" and the magician selects the letter "Z". The wheel is spun by the spectator. (Look at the diagram.) The spectator's letter landed on #36. The magicians letter landed on #49, and he wins. The game can be also changed to whoever has the lower number wins. Game is usually played best of 10 rounds with the players picking different numbers each time.

The Secret: Whatever letter the spectator picks, count 5 letters clockwise (mentally of course!) and this will be your number. For example, if the spectator picks "J", you will pick "M". If the spectator picks "B", you will pick "R". And so one. According to the odds, you will win about 15 times out of every 16 spins. To prevent the spectator figuring out the rule of five system, you can vary the strategy. Count 3,4,10 or 15 spaces. You will still have an advantage, just not quite as large. If you want the person to win , count 1,6,11 or 12 spaces. If the spectator wants you to go first, just pick a number at random. You can then take turns going first, in which case you will almost always win. You should let the other player win now and then to show the game is on the up and up.

Hold'em Proposition Wager

(This article is reprinted from my YouTube Channel BruceMoose77. Mike Caro wrote about this wager a few decades ago in Gambling Times Magazine)

   Which one of these hands would you choose in a heads up Holdem poker game? The hand you pick will be used for the entire game.

You present a spectator with these 3 hands and invite them to choose a hand. And you also tell them that you will pick one of the remaining hands to play against them.

Hand #1...4 Diamonds, 4 Clubs (Pair of 4's.)

Hand #2...J Hearts, 10 Hearts (Jack,Ten suited.)
Hand #3...A Clubs, K Spades (Ace, King unsuited.)

After they choose their hand, and you picked your hand, the hand not chosen is put back into the deck. The deck is now shuffled. Deal eight 5 card Holdem boards with the remaining 48 cards. Compare your hands with each board to see you wins. Keep score. When finished, shuffle all the cards again and repeat the procedure. You will probably need at least the results of 100 Holdem boards. The net result be be that no matter what hand they pick, you can pick a hand that will beat theirs percentage wise overall. For example: 1. They pick Ace-King unsuited, you pick the pair of 4's. You will win about 54% of the time. 2. They pick the pair of 4's, you pick Jack-Ten suited. You will win about 52% of the time. (Due to flush potential.) 3.They pick Jack-Ten suited, you pick Ace-King unsuited. You will win about 58% of the time.(2 over cards potential.) Strange, eh? A beats B beats C beats A.

Note: It is best to find someone with a Texas Holdem hand simulator, or locate one on the internet. That way you can play out about 1000 hands in short order.

Lucky Bingo Cards

(This article is reprinted from my YouTube Channel <u>BruceMoose77</u>**.**

An idea by Donald E. Knuth
Effect: The magician and a spectator each select a bingo card. The numbers 1 through 6 are randomly drawn without replacement, as in standard bingo. If a called number is on a card, it is marked with a coin. The first player to complete a HORIZONTAL row wins. Secret: The game is intransitive. The spectator picks his card first, and then the magician picks a card. According to Martin Gardner, card A beats card B,card B beats Card C, card C beats card D, and card D beats card A! So, A beats B beats C beats D beats A. If, for example,a 12 game match is played, the magician will win the majority of the 12 games. The intransitive nature of the cards is hidden, as opposed to intransitive dice. (On dice, you can see if one die has higher numbers than the other. But these bingo cards, ????) Also, in actual performance, the cards would be secretly marked, so you know which card beats the other. Make sure cards are numbered exactly as shown.

1	2
3	4

A

2	4
5	6

B

1	5
2	6

C

1	3
4	5

D

Red you Win, Black they Lose

(This article is reprinted from my YouTube Channel BruceMoose77. See a video demonstration)

Red Black betting game is similar to the proposition bet " Penny Ante", but uses a deck of playing cards. The Penny game was first presented by Walter Penney in an article, only ten lines long, in the Journal of Recreational Mathematics in 1969. That game involved guessing a sequence of heads and tails in a run of coin flips.

From Wikipedia:

At the start of a game each player decides on their three color sequence for the whole game. For example red red black or black red black. The magician then picks his sequence of cards. Every time the 1st or 2nd player sequence of cards appears, all those cards are removed from the game as a "winning trick" and all cards that have already been turned over are discarded. This continues until the full pack of 52 cards is used. At the end the player with the most "tricks" is declared the winner. An average game will consist of around 7 "tricks". Due to the repeated nature of this game, the second players chance of winning is greatly increased.

To Repeat: Turn the cards over one at a time, placing them in a line, until one of the chosen triples appears.

The winning player takes the upturned cards, having won that trick. The game continues with the rest of the unused cards, with players collecting tricks as their triples come up, until all the cards in the pack have been used. The winner of the game is the player that has won the most tricks.

Example: Spectator picks Black Red Black "BRB", and you then pick Black Black Red "BBR". The cards are dealt in a line off the top of the deck until the winning sequence shows.

Sample sequence: BRRBBR and the sequence BBR appears. The magician wins a trick. These cards are picked up and the deal continues. BRRBRRBRB and the sequence BRB appears and the spectator wins a trick. These cards are picked up and the deal continues. If you run out of cards, you gather all the cards together, shuffle, and start over. First one to win 7 tricks is the winner.

Remember, they always pick first. What a great bar bet! Secret method below.

The Secret Chart: Look at those win percentages for you!!

If they pick BBB, you pick RBB. 99.49 win%

If they pick BBR, you pick RBB, 93.54 win%

If they pick BRB, you pick BBR, 80.11 win %

If they pick BRR, you pick BBR, 88.29 win %

If they pick RBB, you pick RRB, 88,29 win %

If they pick RBR, you pick RRB, 80.11 win %

If they pick RRB, you pick BRR, 93.54 win %

If they pick RRR, you pick BRR, 99.49 win %

If the game ends in a tie, play again.

The original game involved flipping a coin and keeping track of heads and tails until a winning sequence came up. Heads Heads Tails etc, instead of red or black.

The Suit Game

Here is a little known mathematical principle applied to card magic by Martin Gardner. The mathematical principle is called the Gilbreath Principle. Impossible to figure out and no sleight of hand.

Preparation: (Which is done in private.) Remove all the diamonds from the deck and put them aside. The deck is stacked from the top down Spades, Hearts, Clubs, Spades, Hearts, Clubs, and so on.

Put this 39 card deck FACE UP on the table. You can quickly show the cards without giving the order away. Tell the spectator to cut the deck near the middle, if the face card of the LOWER packet is a spade do not complete the cut. The spectator then riffle shuffles the two halves together., the deck is squared, and turned face down. Note: If the spectator does not cut to a spade, have them complete the cut, and cut again. When you see that the LOWER packet has a spade, proceed as above. You do not mention anything about cutting to a spade to the spectator!. Remember, have the spectator keep cutting the deck until they cut to a spade, and THEN have them riffle shuffle the halves together.

Dealing: You tell the spectator that the cards will be dealt off the top of the deck three at a time. But before they do, you introduce the wager. (Instead of money, you can wager beers, potato chips, or whatever. No need to wager money.)

Now, ask the spectator to name 2 of the suits (No diamonds of course.)

If he chooses Clubs and Hearts

You say: For every pair of hearts (or clubs, if he chooses.), I get 25 cents. For every pair of spades, you get $1.00

If he chooses Spades and Hearts

You say: For every pair of spades, I will give you a $1.00. For every pair of hearts, I get 25 cents.

If he chooses Spades and Clubs

You say: For each pair of spades, I will give you a $1.00. For each pair of clubs, I get 25 cents.

Now you start. Say they choose spades, and you have clubs. Deal 3 cards off the top of the deck. If there are no matching suits in the group, toss aside. Continue dealing like this. If there are 2 clubs in the group, put this group in front of you. If there 2 spades in the group, put the cards in front of the spectator. Finish out the deck dealing in this method.

The bet seems fair, but you will always win even if you give them 10000-1 odds. You will have to dress this effect up to suit your style, since it is not a "flashy" type effect.

Two Notes: "Rentacow" on YouTube writes: If you need to pull the trick off with only one cut and she cuts to any suit she didn't choose (this will happen more than half the time so expect it), use a flipped script after she shuffles the deck. Say "I have been having such a bad luck streak lately. I'll bet i don't win a single hand...

"Xavier Quincy" on YouTube writes: So how do you stick the spectator with spades when he picks hearts and clubs? The explanation sounds too contrived. Anyway, it'd be better if you let the spectator select HIS suit and YOUR suit. The trick will work if he picks the suits BEFORE the shuffle. If his suit is spades, cut to a spade (as in the video version) and do the riffle shuffle. If his suit is hearts, cut to a club and shuffle. If his suit is clubs, cut to a heart and shuffle.

Animal Dice

   This is my own invention. The trick is based on Efron dice which are labeled like this:

The four dice A, B, C, D have the following numbers on their six faces:

Die A: 4, 4, 4, 4, 0, 0
Die B: 3, 3, 3, 3, 3, 3
Die C: 6, 6, 2, 2, 2, 2
Die D: 5, 5, 5, 1, 1, 1

Die A beats B 66% of the time.
Die B beats C 66% of the time.
Die C beats D 66% of the time.
Die D beats A 66% of the time.

The Game:
A spectator chooses a die. Say she chooses "B". You will choose "A". If she chooses "D", you will choose "C". You both roll your die simultaneously , whoever has the higher number wins. You play a set match of about 10 rolls. (You each keep the same die for the entire match of course.)

My idea is to substitute animals for the numbers.

Elephant beats Lion
Lion beats monkey
Monkey beats Mongoose
Mongoose beats Snake
Snake beats Mouse
Mouse scares (beats) Elephant
And everybody beats the Fly.

The four dice A, B, C, D have the following animals on their six faces:

Die A: Monkey Monkey Monkey Monkey Fly Fly

Die B: Mongoose Mongoose Mongoose Mongoose Mongoose Mongoose

Die C: Elephant Elephant Snake Snake Snake Snake

Die D: Lion Lion Lion Mouse Mouse Mouse

Idea: Use your computer's graphic capabilities to make a set of animal cards. Make four piles A,B,C,D with the above animals in each pile. They pick a pile,you pick a pile. You each shuffle the cards face down. Then turn the top card face up. Whoever has the stronger animal wins.

James Grime Dice

These 3 dice are numbered like this:

Die A: 3 3 3 3 3 6
Die B: 2 2 2 5 5 5
Die C: 1 4 4 4 4 4

How the game is played: The spectator picks a die first, and then the magician picks a die. Each person rolls their die.. Whoever has a higher number wins. Over the course of 10 rolls or more, Die A beats Die B, Die B beats Die C, Die C beats Die A.

What if the spectator tells you to pick first? You change the rules slightly. You pick a die and he picks a die. Increase the stakes. Each person now rolls his die TWICE and adds the total together. According to Grime, now C beats B beats A beat C!

Cheng's Magic Gambling Die

Martin Gardner stated a student at Bath University in England named R. C. H. Cheng invented this magic die and sent it to him. Gardner published it in the June, 1983 issue of Epoptica magazine.

It is a single large 6 sided die. Each face of the die has the numerals 1 through 6 on it, and each numeral is a different color. The die is set up like this chart.

	Red	Orange	Yellow	Green	Blue	Purple
Side A	1	2	3	4	5	6
Side B	6	1	2	3	4	5
Side C	5	6	1	2	3	4
Side D	4	5	6	1	2	3
Side E	3	4	5	6	1	2
Side F	2	3	4	5	6	1

To read chart:
On side "A", 1 is red, 2 is orange, 3 is yellow, etc.
On side "B", 6 is red, 1 is orange, 2 is yellow, etc.

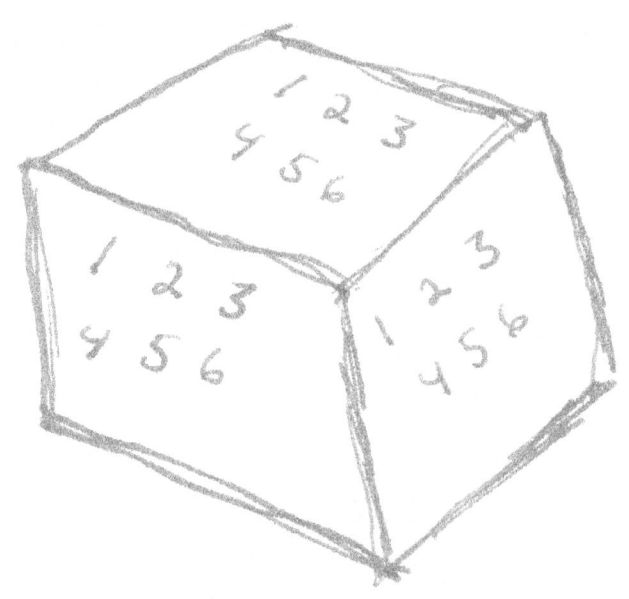

This is how the game is played. The spectator picks a color and the magician picks a color. He rolls the die. The color with the highest number wins. Example: He picks orange and you pick yellow. The die is rolled and say it lands on side "D". Number 5 is orange and number 6 is yellow, and you win. Example #2: The die is rolled and lands on side "F". Number 3 is orange and number 4 is yellow. You win again.

The secret? Whatever color he picks, you pick a color to the right of his in the chart. (And the chart is circular.) For example, if they pick yellow, you pick green. If they pick green, you pick blue. If they pick purple, you pick red. You will win 5 out of 6 times! You can vary the odds by picking 2 colors away, or have the spectator win now and then by picking a color to his left.

Magician Mel Stover suggested using a 6 sided rolling log.

Gardner suggested making this up with colored numerals on cards. Packet "A" is composed of 6 cards corresponding to side "A" on the die, packet "B" is composed of 6 cards corresponding to side "B", and so on. You will end up with 6 packets of cards with 6 cards in each packet. Tell the spectator to select a color, and then pick a packet. Magician selects a color and picks a packet. Each looks through their packet to find their color and see what number it is. High number wins.

Also you might want to make an origami die.

Back to Rock Paper Scissors

Here is a version by magician Max Maven (Phil Goldstein). Get 3 plastic poker chips or small disks of three different colors, and small round blank stickers. For the sake of description you have 3 poker chips: 1 red, 1 white and 1 blue. Put a sticker on each side of the 3 chips. On the white chip draw a rock on one side and paper on the other. On the blue chip draw a rock on one side and scissors on the other. On the red chip draw a scissors on one side and paper on the other.

The game: Spectator chooses a chip and magician chooses a chip. Each person shakes his chip in his hand and tosses it on the table. Rock beats scissors Scissors beats paper. Paper beats rock. Whoever wins the round scores 1 point. Repeat. First person to get 10 points wins. Magician wins most of the time.

The Secret. Remember Red beats white. White beats blue. Blue beats red. Whatever the spectator picks you pick the color that will beat his chip. The magician has a 67% chance to win.

Noah's Ark

Magician introduces a deck of animal cards. This type of deck is found in the children section of a toy store. The deck consists of different animals on each card instead of playing card values or suits. Spectator is asked to pick out 4 different animals and discard the rest of the deck. Say the four cards are a rat, turtle, lion, and bat. Spectator is now asked to name two of the animals. Say she names bat and rat. Magician names lion and bat. The 4 cards are picked up by the spectator and shuffled face down. Cards are turned up in any order. If bat and rat turn up in that order, spectator wins. If lion and bat turn up in that order magician wins. Cards are then picked up, shuffled, and the deal repeated. Best of 20 rounds is the winner. Magician will win 60% of the time.

Examples: 1. bat turtle rat lion. No winner. The rat card must come right after the bat card to win.

2. Lion bat turtle rat. Magician wins.
3. Rat lion turtle bat. No winner.
4. Lion turtle bat rat. Spectator wins.

The secret: This is a variation of Walter Penney's

coin flipping game. When the spectator picks the four animal cards, put them in a certain order in your mind. It can be alphabetical or another easily memorized order.

My order is bat rat turtle lion. (Imagine the order is circular. Bat rat turtle lion bat rat turtle etc.) When the spectator picks her pair of animals, her first animal is your second . Say she picks bat and rat. Your second animal is bat and your first animal is the one before in the memorized order. In this case lion. Your animals are lion bat. Say she picks lion and turtle. Your second animal is lion and your first animal is turtle.(The one before it in the memorized order.) Your animals are turtle lion.

Remember, her first animal is your second animal and your first animal is the one before in your memorized order.

Word Bingo

   This appears related to Donald Knuth's Bingo game. You need 5 bingo cards and 5 bingo "balls." The bingo cards have letter pairs on them. (See diagram) The bingo "balls" are 5 slips of paper with one letter from the word "Bingo" on them. Make these up and you are ready to go.

   To play the game: The spectator goes first and can choose any bingo card. The magician now selects his card. The five bingo balls are turned face down and mixed. One is turned face face up and the players mark the cards accordingly. After the slip is turned up and used, it is put aside. First player to cover a word pair wins. Magician wins 60% of the time.

The secret: Look at the bingo cards. Each card has one letter that is repeated twice. One card has 2 B's, another 2 I's, 2 N's, 2 G's, and 2 O's. If the spectator picks a card with 2 B's, you pick one with 2 I's. If they pick one with 2 G's, you pick the one with 2 O's. If they pick the one with 2 O's, you pick the card with 2 B's. (Just remember the order of cards in BINGO!. B beats I beats N beats G beats O beats B beats I, etc.) When presenting the effect, make sure the bingo cards are scrambled so as to hide the double letters.

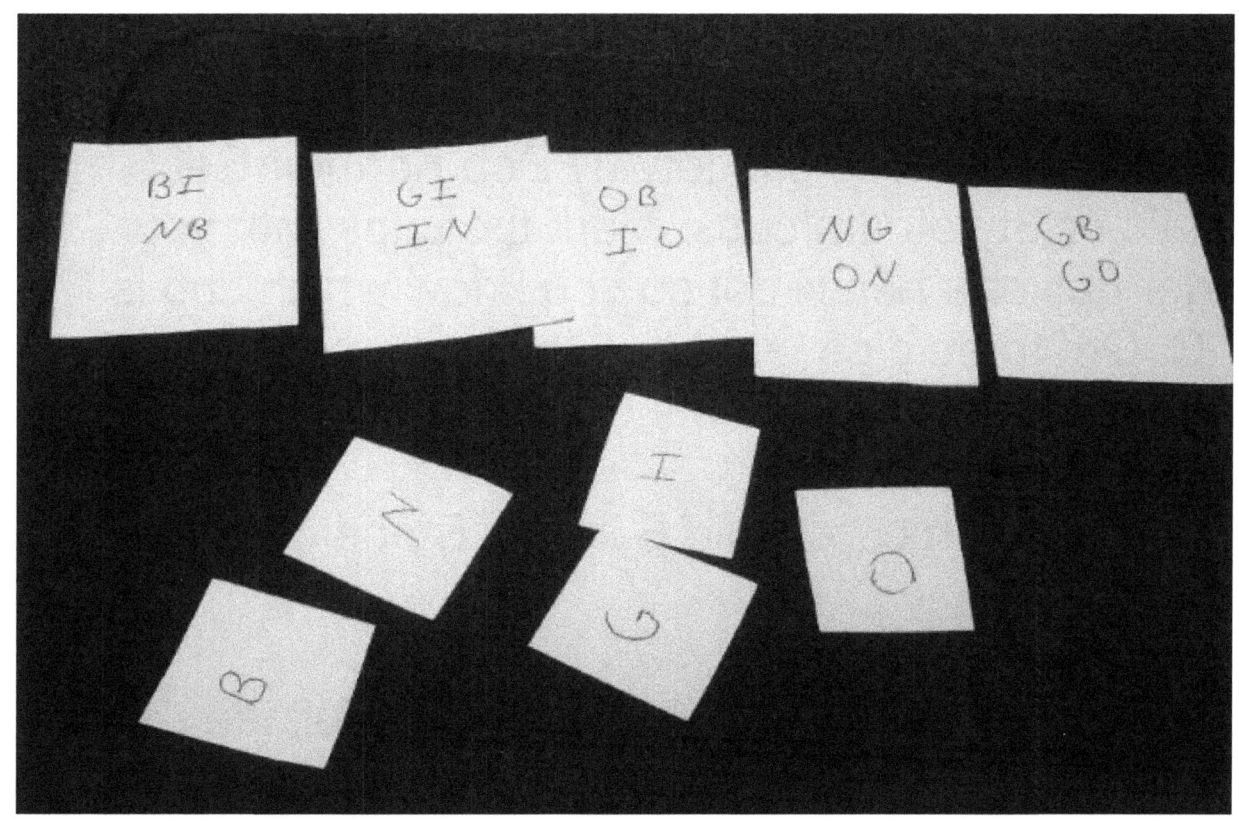

Under the Cups

Here is a ingenious subtlety by Bob Farmer. Get 3 plastic or Styrofoam cups. Mark the cups with your finger nail or a pencil dot so you know which cup is A,B or C. Get three dice that have the non transitive property of A beats B beats C beats A.

1. Put die A in cup A, die B in cup B and die C in cup C.

2. You walk away or cover your eyes. Tell the spectator to shake a cup and turn it face down. Tell him not to peek at the result. Do the same with the other 2 cups.

3. There are now 3 cups face down with a die under each. No one knows what the top number is of the 3 dies.

4. The spectator picks a cup and you pick a cup. The cups are lifted at the top numbers compared. High number wins. You will win 60% of the time.

Drawing the Short Straw

A Karl Fulves idea. Once again get 3 plastic or Styrofoam cups. And again mark the cups with your finger nail or a pencil dot so you know which cup is A,B or C. Get a couple of straws and cut the into 9 different lengths. Either 1 centimeter, 2 centimeters, up to 9 centimeters. Or 1 inch, 2 inches, 3 inches, up to 9 inches. Cut 3 small holes in the top of each cup. Place the straws in each cup as follows. (See diagram)

Cup A. 2 centimeters (or inches), 3 centimeters, 9 centimeters

Cup B. 1 centimeter, 7 centimeters, 8 centimeters

Cup C. 4 centimeters, 5 centimeters, 6 centimeters

Cup A beats B. B beats C. C beats A. Spectator picks a cup and pulls a straw from that cup. You pick a cup and pull a straw from that cup. Compare straws. Longer straw wins. You will win 63% of the time. Replace straws, scramble cups, and draw again.

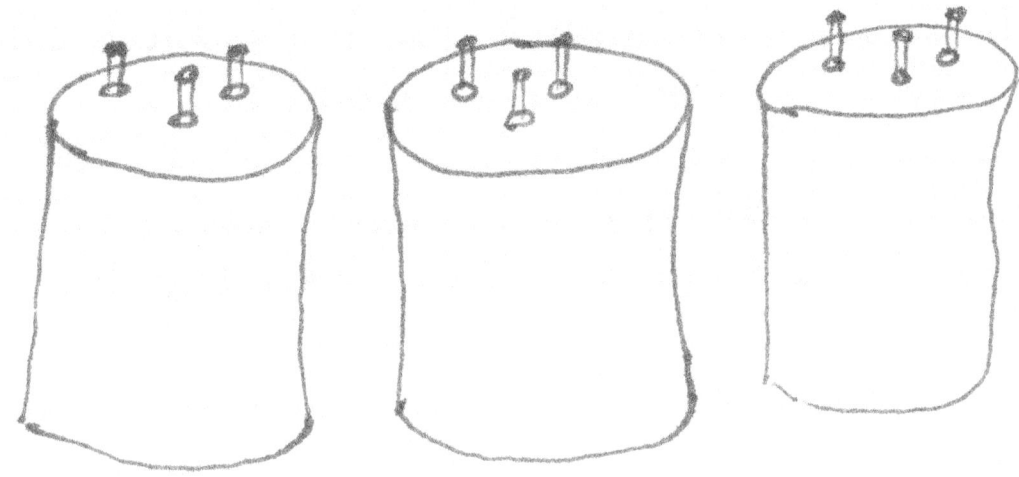

Bibliography

Karl Fulves. "<u>SWINDLE & CHEAT</u>" 1991,71 pages. This is the ONLY and best book dedicated to non-transitive games and magic. Very hard to find.

Colm Mulcahy at Cardcolm.org His Card Colm Column has as an article "Dicey Cards" on non transitive games.

Martin Gardner. In his book "<u>Wheels, Life, and other Mathematical Amusements</u>" has article on "The Paradox of Non-transitive Dice." In the book "<u>Time Travel and other Mathematical Bewilderments</u>" has a great chapter on some odd non-transitive games, which I put into the book you are holding.

Bob Farmer has some articles in an old column called Flim - Flam in Magic magazine. See "Bar Racuda" in April 2000 issue, "Six Bones of the Beast" in May 2000 issue, and "Uncle Sam Still Wants You" in August 2000 issue. Farmer also recommends using dice with cards suits on them to disguise the numbering.

Wikipedia has an introductory article.

James Grime website. Has article on how to defeat 2 opponents at the same time! Http://www.singingbanana.com/dice/article.htm . Also you can buy some dice here.

Grand Illusions.com has non transitive dice.

Ebay sometimes has non transitive dice for sale.

Various obscure mathematical papers can be found on the internet. But these are usually VERY mathematical as opposed to entertaining.

If you write to some of the magic forums, they can point you to some of the obscure magic publications that contain tricks and articles on non-transitive games and magic.

Non transitive dice for Sale

(Grime Dice) http://amzn.to/1ITTK4F

(Grand Illusion Dice Set of 4) http://amzn.to/1dUsJAY

(Grime dice/small set) http://amzn.to/1O40ih2

Videos

Grime Dice https://www.youtube.com/watch?v=zWUrwhaqq_c

Sicherman & Non transitive Dice https://www.youtube.com/watch?v=zWUrwhaqq_c
Scam School https://www.youtube.com/watch?v=056TWryEkoQ

Custom Dice

Make your own dice using Makerbot 3D printer
http://www.thingiverse.com/thing:76772

You can have custom dice made at:

Shape ways http://www.shapeways.com/create?li=nav

Chessex http://goo.gl/UmCJjG

Q-workshop http://q-workshop.com/customdice

Crystalcaste http://crystalcaste.com/custom.htm

Game Crater https://www.thegamecrafter.com/custom-dice

And last but not least..If you have any corrections to the book, any ideas for effects, criticisms or praise, please send them to:
UNDERGROUNDMAGIC@HUSHMAIL.COM

www.ingramcontent.com/pod-product-compliance
Lightning Source LLC
Chambersburg PA
CBHW080653180526
45168CB00008B/3405